PLANT PARTS
WE EAT

ROOTS WE EAT

Katherine Rawson

Creating Young Nonfiction Readers

EZ Readers lets children delve into nonfiction at beginning reading levels. Young readers are introduced to new concepts, facts, ideas, and vocabulary.

Tips for Reading Nonfiction with Beginning Readers

Talk about Nonfiction

Begin by explaining that nonfiction books give us information that is true. The book will be organized around a specific topic or idea, and we may learn new facts through reading.

Look at the Parts

Most nonfiction books have helpful features. Our *EZ Readers* include a Contents page, an index, and color photographs. Share the purpose of these features with your reader.

Contents

Located at the front of a book, the Contents displays a list of the big ideas within the book and where to find them.

Index

An index is an alphabetical list of topics and the page numbers where they are found.

Photos/Charts

A lot of information can be found by "reading" the charts and photos found within nonfiction text. Help your reader learn more about the different ways information can be displayed.

With a little help and guidance about reading nonfiction, you can feel good about introducing a young reader to the world of *EZ Readers* nonfiction books.

Mitchell Lane
PUBLISHERS

2001 SW 31st Avenue
Hallandale, FL 33009
www.mitchelllane.com

First Edition, 2021.

Author: Katherine Rawson
Designer: Ed Morgan
Editor: Morgan Brody

Names/credits:
Title: Roots We Eat / by Katherine Rawson
Description: Hallandale, FL :
Mitchell Lane Publishers, [2021]

Series: Plant Parts We Eat
Library bound ISBN: 978-1-58415-064-0
eBook ISBN: 978-1-58415-065-7

EZ Readers is an imprint of Mitchell Lane Publishers.

Photo credits: Freepik.com, Shutterstock

Contents

Plants have roots.
The roots are underground.
They hold the plants in the **soil**.

Roots help the plant grow.
They take in water from the soil.
They take in **nutrients**.

Some roots are good to eat.
Carrots are roots.
They are sweet and **crunchy**.

8

DID YOU KNOW?

The orange color of carrots comes from a nutrient called beta carotene. It helps our eyes, bones, teeth, and skin stay healthy.

Some carrots are long and thin. Some are short and fat. Carrots can be orange, yellow, purple, or even white!

11

Beets are roots.
They are round and fat.
They can be red or orange.

DID YOU KNOW?

Candy cane beets are plain red on the outside. On the inside, they are striped like a candy cane.

We can **grate** them or **slice** them.
They taste good in salads and soups.

DID YOU KNOW?

Watermelon radishes are pink on the inside, just like a watermelon.

Radishes are roots.
They come in all sizes.
Some are as small as a quarter.
Some are as long as your arm.

Radishes are **spicy**.
They taste good in salads
and sandwiches.
We can cook them, too.

DID YOU KNOW?
The smallest radishes are less than an inch wide.

Sweet Potato

Daikon Radish

Did You Know?
These long radishes look like white carrots.
They can grow to be 20 inches long.

PARSNIP

TURNIP

Long, short, or round, roots are good to eat!

21

Glossary

crunchy
Not soft or mushy

grate
Break into small pieces by rubbing against
something rough

nutrients
Contents of food that we need for good health

slice
Cut into thin pieces

soil
Dirt in which plants grow

spicy
Having a hot flavor

Sources

http://www.justscience.in/articles/structure-and-function-of-plant-roots/2017/07/01

https://www.consumerreports.org/healthy-eating/are-beets-good-for-you/

http://www.berkeleywellness.com/healthy-eating/food/article/types-beets

https://www.fruitsandveggiesmorematters.org/baby-candy-cane-beets-nutrition-selection-storage

https://garden.org/learn/articles/view/604/

http://www.berkeleywellness.com/healthy-eating/food/article/types-carrots

http://www.carrotmuseum.co.uk/betacarotene.html

http://www.berkeleywellness.com/healthy-eating/food/article/types-radishes

Further Reading

Web Pages:
Read more about roots:
https://extension.illinois.edu/gpe/case1/c1facts2a.html

Read more about plant parts:
https://www.dkfindout.com/us/animals-and-nature/plants/parts-plant/

Books:
The Amazing Life Cycle of Plants
by Kay Barnham
(B.E.S. Publishing, 2018)

Seed to Plant
by Kristin Baird Rattini
(National Geographic Children's Books, 2014)

Index

About the Author

Katherine Rawson loves growing, cooking, and eating vegetables. She also loves writing. So, she thought it would be a great idea to write books about plants we eat. Her favorite root vegetable is a purple carrot.